SAFE MANUAL HANDLING

A Gower Health and Safety Workbook

Graham Roberts-Phelps

Gower

Published by
Gower Publishing Limited
Gower House
Croft Road
Aldershot
Hampshire GU11 3HR
England

Gower
Old Post Road
Brookfield
Vermont 05036
USA

British Library Cataloguing in Publication Data
Roberts-Phelps, Graham
Safe manual handling. – (A Gower health and safety workbook)
1.Lifting and carrying – Safety measures
I.Title
363.1′1′9658781

ISBN 0 566 08062 1

Typeset in Times by Wearset, Boldon, Tyne and Wear and printed in Great Britain by print in black, Midsomer Norton.

Contents

Chapter 1 Introduction

This first chapter acts as a record of your progress through the workbook and provides a place to summarize your notes and ideas on applying or implementing any of the points covered.

PERSONAL DETAILS

Name:	
Position:	
Location:	
Date started:	Date completed:

Chapter title	Signed	Date
1. Introduction		
2. Understanding safe manual handling		
3. Preventing manual handling injury		
4. Six steps to safe lifting		
5. Look after your back		
6. Learning review		
Demonstration of safety in the workplace		
Steps taken to reduce risks and hazards		

Lifting safety review dates	Assessed by	Date
1 month	______________	______________
2 months	______________	______________
3 months	______________	______________
6 months	______________	______________

HOW TO USE THIS SELF-STUDY WORKBOOK

Overview

This self-study workbook is designed to be either one, or a combination, of the following:

- a self-study workbook to be completed during working hours in the student's normal place of work, with a review by a trainer, manager or safety officer at a later date

- a training programme workbook that can be either fully or partly completed during a training event or events, with the uncompleted sections finished in the student's normal working hours away from the training room.

It contains six self-contained chapters which should each take about 20 minutes to complete, with the final section, 'Learning Review', taking slightly longer due to the testing and validation instruments.

It is essential that students discuss their notes and answers from all sections with a supervisor, trainer or coach.

NOTES FOR TRAINERS AND MANAGERS

For use in a training session

If you are using the workbook in a training event you might choose to send the manual to students in advance of their attendance, asking them to complete the Introduction (Chapter 1). Other exercises can then be utilized as required during the programme.

For use as an open-learning or self-study tool

Make sure that you have read the workbook properly yourself and know what the answers are. Anticipate any areas where students may require further support or clarification.

Comprehension testing

Each section features one or more summary exercises to aid understanding and test retention. The final chapter, 'Learning Review', contains a set of tests, case studies and exercises that test application and knowledge. Suggested answers to these are given in the Appendix.

If you are sending the workbook out to trainees, it is advisable to send an accompanying note reproducing, or drawing attention to, the points contained in the section 'Notes for Students'. Also, be sure to set a time deadline for completing the workbook, perhaps setting a review date in advance.

The tests contained in the learning review can be marked and scored as a percentage if required. You might choose to set a 'pass' or 'fail' standard for completion of the workbook, with certification for all those attaining a suitable standard. Trainees who do not reach the required grade on first completion might then be further coached and have points discussed on an individual basis.

NOTES FOR STUDENTS

This self-study workbook is designed to help you better understand and apply the topic of safe manual handling. It may be used either as part of a training programme, or for self-study at your normal place of work, or as a combination of both.

Here are some guidelines on how to gain the most from this workbook.

- Find 20 minutes during which you will not be disturbed.
- Read, complete and review one chapter at a time.
- Do not rush any chapter or exercise – time taken now will pay dividends later.
- Complete each written exercise as fully as you can.
- Make notes of questions or points that come to mind when reading through the sections.
- Discuss anything that you do not understand with your manager, safety officer or work colleagues.

The final chapter, 'Learning Review', is a scored test that may carry a pass or fail mark.

At regular intervals throughout the workbook there are exercises to complete and opportunities to make notes on any topics or points that you feel are particularly important or relevant to you. These are marked as:

Notes

LEARNING DIARY

Personal Learning Diary

Name: ________________________________

Job Title: ________________________________

Company: ________________________________

Date: ________________________________

The value of the training programme will be greatly enhanced if you complete and review the following Learning Diary before, during and after reviewing and reading the workbook.

Learning Objectives

At the start or before completing the workbook, please take time to consider what you would like to learn or be able to do better as a result of the training process. Please be as specific as possible, relating points directly to the requirements of your job or work situation. If possible, please involve your manager, supervisor or team leader in agreeing these objectives.

Record these objectives below

1.

2.

3.

4.

5.

6.

Please complete before continuing

Learning Log

During the training programme there will be many useful ideas and learning points that you will want to apply in the workplace.

Key ideas/learning points	How I will apply these at work

Please complete before continuing

LEARNING APPLICATION

As you complete each chapter, please consider and identify the specific opportunities for applying the skills, knowledge, behaviours and attitudes and record these below.

Action planned, with dates	Review/comments

Remember, it may take time and practice to achieve new results. Review these goals periodically and discuss with your manager.

PLEASE COMPLETE BEFORE CONTINUING

HOW TO GET THE BEST RESULTS FROM THIS WORKBOOK

The format of this workbook is interactive; it requires you to complete various written exercises. This aids both learning retention and comprehension and, most importantly, acts as a permanent record of completion and learning. It is therefore essential that you **complete all exercises, assignments and questions**.

In order to gain the maximum value and benefit from the time that you invest working in completing this workbook, use the following guidelines.

Pace yourself

You might choose to work through the whole workbook in one session or, alternatively, you might find it easier to take one chapter at a time. This is the recommended approach. If you are using this workbook as part of a live training programme, then make time to follow through any unfinished exercises or topics afterwards.

Share your own opinions and experience

We all have a different view of the world, and we all have different backgrounds and experiences. As you complete the workbook it is essential that you relate the learning points directly to your own situation, beliefs and work environment.

Please complete the exercises using relevant examples that are personal and specific to you.

Keep an open mind

Some of the material you will be covering may be simple common sense, and some of it will be familiar to you. Other ideas may not be so familiar, so it pays to keep an open mind, as most learning involves some form of change. This may take the form of changing your ideas, changing an attitude, changing your perception of what is true, or changing your behaviours and the way you do things.

When we experience change, in almost anything, our first automatic reaction is resistance, but this is not usually the most useful response. Remember, safety is something we have been aware of for a long time, and consider (or fail to consider, as the case may be!) every day. As a result, we follow procedures without thinking – on auto-pilot as it were. This often means that we have a number of bad habits of which we are unaware.

Example of change:

Sign your name here as you would normally do:

Now hold the pen or pencil in the ***opposite*** *hand to that which you normally use and sign your name again:*

Apart from noting how difficult this might have been, consider also how 'strange' and uncomfortable this seemed. You could easily learn to sign your name with either hand, usually far more quickly than you might think. However the resistance to change may take longer to overcome.

Making notes not only gives you information to refer to later, perhaps as a review of the workbook, but it also aids memory. Many people find that making notes actually helps them to remember things more accurately and for longer. So, as you find things that are particularly useful or of particular interest, please take a couple of moments to write these down, underline them or make comments in the margin or spaces provided.

Review with others

In particular, ask questions and discuss your answers and thoughts with your colleagues and fellow managers, especially points which you are not sure of, points which you are not quite clear on, and perhaps points about which you would like to understand more.

Before you start any of the main chapters, please complete the following learning assignments.

LEARNING OBJECTIVES

It is often said that if you do not know where you are going, any road will get you there. To put it another way, it is difficult to hit the target you cannot see. To gain the most benefit from this workbook, it is best to have some objectives.

Overall objectives

- **Improvements.** We don't have to be ill to improve our fitness. Improvement is always possible.
- **Skills.** Learn new skills, tips and techniques.
- **Knowledge.** Gain a better understanding of safety issues.
- **Attitudes.** Change the way we think about safety issues.
- **Changes.** Change specific attitudes on behaviours and our safety procedures and practice.
- **Ideas.** Share ideas.

Here are some areas in which you can apply your overall objectives.

1. Hazards and risks

The first objective is to be able to identify safety hazards and risks. These may exist all around us and may not be readily identifiable as such – for example, the ordinary moving of boxes or small items, using a kettle or hand drill, cleaning and so on.

2. Prevention

Prevention is always better than cure, and part of this workbook will deal with knowing how to prevent accidents and injuries in the first place. Injuries are nearly always painful both in human and business terms. As well as accidents that cause us or others harm, there are many more accidents that cause damage and cost money to put right.

3. Understanding your safety responsibilities

Health and safety is everybody's responsibility, and safety is a full-time job. As you complete this workbook you will be looking at how it affects you personally and the role that you can play, not only for your own safety but also for the safety of others around you.

4. Identifying ways to make your workplace safer

A workbook like this also gives us the opportunity to put ideas together on how we can improve the health and safety environment of our workplace. We do not have to have safety problems in order to improve safety, any more than we have to be ill to become fitter.

An improvement in working conditions does not have to cost much or be very complicated.

Make a note here of any personal objectives that you may have.

Notes

Lifting Quick Check

1. List five things that you have lifted, carried or moved today.	1. 2. 3. 4. 5.
2. List three additional lifting operations that you would expect to perform in the course of a normal day or week.	1. 2. 3.
3. Rate each as high, medium or low level of risk.	

Please complete before continuing

HOW SAFE IS YOUR JOB?

Take time to reflect on these questions, perhaps making some notes. Consider not only your experience in your current job, but in previous ones as well.

1. *Have there been any manual handling or lifting-related accidents or injuries in your workplace, or have you experienced any personally?*	
2. *If so, what do you think might have caused them or been a contributing factor?*	
3. *Have you ever suffered backaches or injuries and what do you think may have caused these?*	

PLEASE COMPLETE BEFORE CONTINUING

Chapter 2 Understanding Safe Manual Handling

This chapter emphasizes the importance of safe manual handling, helps you identify high-risk activities and shows you how to prevent injury.

Before starting this chapter, please take a few moments to make a note of any ideas or actions in the learning diary and log in Chapter 1.

Key points to remember:

- ***Understand why safe manual handling is important***
- ***Identify high-risk activities***

HIGH-RISK ACTIVITIES

Here are some of the common hazards that carry higher levels of risk when manual handling.

Traditional lifting

The first high-risk activity is traditional lifting – whether this is shifting boxes, putting things on shelves and loading vans for transport, or just moving things from one place to another in our daily work routine.

Repetitive lifting

The next activity which carries a high risk of causing injury or a manual handling accident is lifting using repetitive or restrictive movements. Although this may often involve very light lifting movements, because there are so many of them, or because of the physical restrictions involved, it can cause cumulative strain injuries. People at risk from this type of injury are often factory workers, shop workers scanning prices at tills and others doing similar repetitive tasks.

Awkward lifting

Another common form of high-risk activity is moving awkward or particularly heavy loads. Whilst these types of manual handling operations often attract more attention and more thought thereby minimizing or reducing the risk, lifting very heavy objects, or those that are difficult to grip, present particular hazards.

Bad posture when lifting

Another frequently overlooked factor that can increase the risk of injury from any lifting activity is working in an awkward posture. Whether this is because of our own ignorance, or difficult working conditions, the result is the same – an increased risk of manual handling accident or injury. Frequent bending is also a problem. This puts undue strain on joints and the back and can lead to very painful back problems and disorders.

Forced or rapid movements

The final high-risk factor involved in manual handling is forced, repetitive or rapid movements. This is common in jobs which entail operating machinery or performing the same task repeatedly. Whilst the task in itself may not be particularly stressful or put undue strain on the body, over time it can lead to chronic injury or pain.

Review the previous section and make some notes on the points which concern you the most.

- Traditional methods of lifting
- Repetitive or restricted movements
- Moving awkward or heavy loads
- Working in an awkward posture
- Frequent bending
- Forceful, repetitive or rapid lifts

Notes

Lifting Analysis

Mark the following types of lifting as either T – traditional, R – repetitive, A – awkward, B – bad posture, F – forced movements:

Carrying a small child up a flight of stairs ☐

Carrying boxes of copy paper into the office ☐

Carrying equipment up a flight of stairs ☐

Lifting and moving planks and scaffolding ☐

Lifting or helping someone out of a chair ☐

Lifting sacks of cement out of a car boot ☐

Loading a large item on to a high shelf ☐

Operating a till checkout ☐

Stacking shelves ☐

Unloading a large consignment of boxes and cartons ☐

Unloading tools and equipment ☐

Please complete before continuing

THE MANUAL HANDLING OPERATIONS REGULATIONS 1992

On any one working day, there are over 90 000 people off work with back problems.

The Manual Handling Operations Regulations were introduced in 1992 as part of an EC directive on Health and Safety. The Workplace (Health, Safety and Welfare) Regulations 1992 also put some emphasis on manual handling operations. The main requirements of these Regulations are as follows.

1. Provide information and training

All employees must be aware of the common hazards that are associated with manually moving loads and frequent forced or awkward movements of the body. They must understand how these can lead, for example, to back injuries and other injuries to the hands, wrists, arms, legs and neck.

2. Maintain safe practice

Employees should be trained on how to lift safely, as well as how to use any lifting or manual handling equipment or facilities provided. These include hoists, trolleys, trucks, steps and so on.

3. Eliminate lifting where possible

It is also important to eliminate manual handling wherever possible. The Regulations state that you must avoid manual handling if a safer method – for example, mechanized lifting – is practicable. This may mean designing jobs to fit the work to the person rather than the person to the work, taking into account human capabilities and limitations and improving efficiency as well as safety. An organization must avoid manual handling, therefore, wherever there is a risk of injury. Any hazardous manual handling operation that cannot be avoided must be properly assessed for risk of injury. Therefore employees should not be asked to lift heavy or awkward objects if:

- they have not been trained
- loads are above the safe limits and therefore represent an unreasonable level of risk of injury.

4. Provide lifting aids

Wherever possible, equipment and lifting aids should be provided, and this equipment should be tested and safe for the use intended. People must be trained in its usage, and it must be regularly maintained.

5. If you cannot eliminate, reduce where you can

Where manual handling cannot be eliminated, it must be reduced. Actions to consider might include:

- providing mechanical help such as a sack truck or hoist
- making loads smaller, lighter or easier to grasp
- changing the system of work to reduce the effort required
- improving the layout of the workplace to make the work more efficient.

Protective clothing may also be needed to protect parts of the body, such as hands and feet, when lifting.

6. Make manual handling safe

All manual handling operations must be assessed for risk, and safe manual handling procedures must be properly documented. One particular hazard that may need special attention is that of repetitive handling, since repeated or awkward movements which are either too forceful, too fast or are carried out for too long can lead to disorders of the arms, hands or legs.

Occupations such as typing, working on a till or scanner, assembly work and so on are particularly hazardous in this respect. The assessment should investigate the gripping, squeezing or pressure required, awkward hand or arm movements, repeated continuous movements, and their speed, as well as the level of intensity and breaks afforded to the worker.

Make a note of any points from this section that concern you.

Notes

Manual Handling Regulations

With regard to the legal requirements, what immediate actions can you identify for improving the standard of manual handling that you do?

Rate your organization on a scale of 1 (poor) to 5 (excellent) for each of the following:

- Information and training ☐
- Enforcing safe practice ☐
- Removing the need for manual handling ☐
- Providing lifting aids ☐
- Reducing manual handling to a minimum ☐
- Lifting assessments ☐

PLEASE COMPLETE BEFORE CONTINUING

SAFE MANUAL HANDLING: THE FACTS

It is estimated that over 5.5 million working days are lost in Britain every year directly as a result of back, or manual handling-related, injuries or disorders. Back pain is also one of the most painful of all muscular injuries, and one from which your body may never fully recover. Once incurred, back injuries and back pain often stay with you for life.

Health and Safety Executive (HSE) studies and research also reveal that over 30 per cent of all injuries in the workplace are caused by, or are related to, lifting, handling or load-carrying in some form. It is not only those people that carry heavy weights or have to lift frequently in their job who are at risk. Manual handling-related injury disorders are common among all types of jobs and employees.

So, whether you work on a building site or in a shop, you are probably equally at risk in some way. The risk has little to do with the weight that you lift.

The most common accidents and their causes

The following information will illustrate the most common types of accident and their associated causes and, hopefully, indicate how employers can use their own accident and ill-health records to see where changes can be implemented.

Figures from the Health and Safety Commission's *HSC Annual Statistics Report* show that, in terms of percentage of all accidents, the most commonly occurring accidents to employees reported to enforcement listings under RIDDOR are:

1.	slips, trips or falls (on the same level)	35%
2.	falls from height	21%
3.	injuries from moving (falling/flying) objects	12%

The three most common accidents to self-employed workers are the same as above although their prevalence is differently distributed:

1.	falls from height	45%
2.	slips, trips or falls (on the same level)	15%
3.	injuries from moving (falling/flying) objects	14%

All these figures include fatalities, major injuries and injuries resulting in three or more days' absence from work. Statistics are for the most recent period available. However, the HSC comment that 'whilst most other accidents stayed relatively unchanged, slip, trip or fall accidents have increased from 26% to 35% for employees for the period 1986 to 1996'.

As can be seen from the employee figures above, the most common injuries are musculo-skeletal injuries (that is, those from manual handling operations), particularly those affecting the back. Indeed, the HSE's statistics on self-reported work-related illness confirm that these types of injury are the most common for both manual and non-manual occupations.

Estimates suggest that around 3.6 million working days per year are lost as a result from back injuries sustained in the workplace. The number of working days lost as a result of falls from a height and machinery-related injuries are thought to be around 1.7 million and 0.75 million respectively.

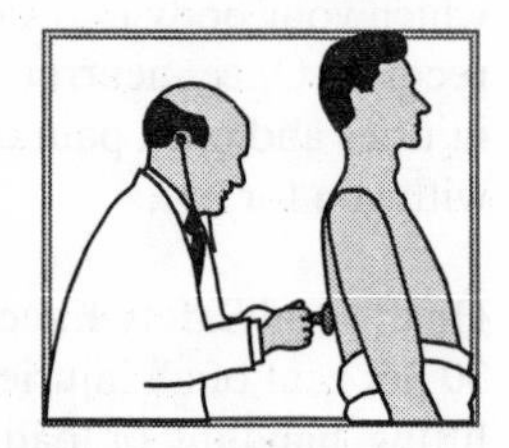

50 per cent of all manual handling accidents are back-related

In manual occupations the next most common work-related illnesses reported were the long-term consequences of trauma and poisoning (which were the biggest causes of workers changing jobs), lung diseases, including asthma, and deafness. In non-manual occupations musculo-skeletal disorders were followed by stress and depression as the second most common complaint and, in offices specifically, headache and eye strain.

Although these particular figures are based on self-reported work-related illnesses – that is, the individuals' perception of their own illness and the relationship with their work – they are broadly in line with other findings.

Make a note of any points from this section that concern you.

Notes

ACCIDENT STATISTICS

Take time to find out the answers to the following questions:

1. *How many accidents, injuries and days off are caused annually by manual handling-related incidents in your organization?*	
2. *Which type of lifting injury is most common?*	
3. *What type of manual handling hazard or accident is most common?*	

PLEASE COMPLETE BEFORE CONTINUING

Chapter 3
Preventing Manual Handling Injury

This chapter provides a six-step approach to preventing manual handling and back injuries.

Before starting this chapter, please take a few moments to make a note of any ideas or actions in the learning diary and log in Chapter 1.

PREVENTING INJURIES

Notwithstanding the Manual Handling Regulations, it makes good sense to lift, carry and move items and equipment with as much care and consideration for your body as possible. Back injuries are both painful and permanent. Once damaged, the back may take years to heal and may never fully recover. Here are six steps to preventing manual handling and back injuries.

Step 1: Look

Take a good look at what you are going to lift and everything around you.

Step 2: Plan

Think through how you are going to lift or carry and assess whether you are going to do it correctly.

Step 3: Grip

Make sure that you are holding the object properly and securely.

Step 4: Think

Consider the activity from a manual handling safety viewpoint.

Step 5: Change

Make sure that you have sufficient breaks and changes of activity within your routine to avoid overloading or overtiring your muscles or back.

Step 6: Ask

Feel able to ask for help advice or assistance.

These are the six steps to preventing and minimizing manual handling accidents and injury when manual handling – that is, loading, carrying, lifting or shifting

Safe manual handling step 1: LOOK

Before even considering lifting an object, or beginning to lift an object, look carefully at what you are going to lift. Ask yourself:

- Where you are actually going to be lifting it to?
- What route or passageway will you be using?
- Does the object have any sharp edges or corners?

- Are there any signs, labels or markings on the outside of the packaging that you should read, particularly regarding weight, contents or which way up the item should be?

Check whether the object which you are moving has any broken or torn parts. Make sure also that you can get a good grip on the object – is it, for example, wet or slippery?

A moment's observation can prevent a great deal of later discomfort.

Safe manual handling step 2: PLAN

Having examined the object that you are going to lift or move, the next stage is to think through the movements you are going to use and the route you are going to take.

However, before you do that, you must consider whether the lifting or the manual handling operation can be avoided or mechanized.

The Manual Handling Operations Regulations of 1992, which is the actual legislation that has been passed to maintain the manual handling safety of manual handling operations in the workplace, state that you must avoid manual handling if a safer method – for example, mechanical – is practicable. Jobs should be designed to fit the work to the person rather than the person to the work. This will take into account human capabilities and limitations and improve efficiency as well as manual handling safety. So think about whether you need to do the job at all, and if there is an easier or safer way.

Once you have considered this, consider the actual mechanics of the lifting or the handling itself.

The principal factors which you may want to consider are as follows:

- How are you going to stand?
- If you need to access a high area, is the floor space clear?
- Is there anything in your way?
- Can you see clearly where you are going?

It is essential to use common sense; never move or lift objects without forethought.

Safe manual handling step 3: GRIP

Getting a good grip is essential. When we start to lose our grip or have failed to gain a proper hold of an object, we will be tempted to strain in order to maintain that grip. This can put undue pressure or strain on joints, muscles and ligaments, and is a frequent cause of manual handling accidents. Grip also stands for **G**et **R**ight **I**n **P**osition. Getting a secure hold on an object is not only a function of your actual physical grasp, but also of how you stand in relation to the object that you are moving or lifting, and of your position over the object. A 'hook' grip, with fingers bent, is better than grasping with your fingers straight.

Consider what is a good grip.

- Can you get your hands under or around the object?
- Can you hold it securely?
- Is the packaging or the outside of the object secure?
- Do you need special gloves?

Safe manual handling step 4: THINK

Always consider ways of actually helping yourself when manual handling or lifting. Assistance can take the form of either a lifting aid, such as a truck, trolley, hoist and so on, or else someone else to physically help you.

Is there any particular protective equipment that would make the operation safer? This might be gloves, shoes, protective clothing, or maybe even a hard hat or protective glasses.

Make a list of all the types of protective and lifting equipment that you know to be available to use in your workplace. Consider for a moment whether you have been trained to use it properly.

Safe manual handling step 5: CHANGE

As mentioned earlier, certain types of lifting or handling constitute what is known as repetitive handling. These are repeated or awkward movements which are forceful, fast or carried out for too long and can lead to disorders of the arms, the hands or the neck. These are common risks for people who operate machinery such as computers, manufacturing items, tills, punches, presses and so on.

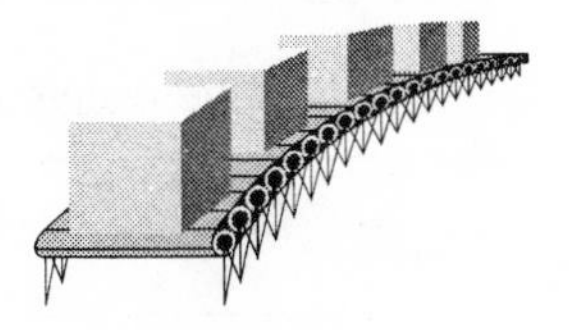

We can reduce or prevent the risks of repetitive handling by:

- reducing the levels of force required, maintaining equipment and using tools with well designed handles, or perhaps changing equipment
- reducing repetitive movements by perhaps varying tasks, rotating jobs, using power-driven tools, reducing the pace or speed of machines and introducing rest and recovery breaks
- eliminating awkward positions by changing the work station or work.

Consider whether there are any cumulative stress or strain hazards in your work and what they mean, or what the term means to you.

The danger of accumulative strain injury is that its effects can go completely unnoticed for many years, and the symptoms are often disguised as other general aches and pains. This has led to some controversy within manual handling safety circles, as it is often very difficult in some cases to relate such injury directly to a work cause.

Safe manual handling step 6: ASK

Asking for advice or assistance is actually a sign of strength, not weakness. If you are not sure about something or need advice, you must ask. Not only is this common sense, it is also a legal obligation under various regulations that if you are unsure about your capabilities or about the techniques of handling or lifting any item, you have a responsibility to ask or question your manager or supervisor about this.

Make a note of any points from this section that concern you.

Notes

Assessment

Please make some notes on the following.

When assessing objects to be lifted or moved what should you be looking for?	
What information might be contained on the box or package?	
Make a list of all the different types of lifting equipment or aids that you could use.	
What type of repetitive movement or lifting occurs in your work?	

Please complete before continuing

SELF-ASSESSMENT WORKSHEET

Please complete the following questionnaire, as honestly and accurately as you can. Rate your response to each statement or question on the following scale.

1 = Never; 2 = Sometimes; 3 = Usually; 4 = Often; 5 = Always.

1. I always consider avoiding manual handling wherever possible	1 2 3 4 5
2. If I see someone lifting, carrying or loading in a way that is *not* particularly manual handling safety-aware, I will point it out to them	1 2 3 4 5
3. I assess the risk of each handling operation that can't be avoided	1 2 3 4 5
4. I use mechanical equipment, such as trucks, hoists, trolleys and so on, wherever possible	1 2 3 4 5
5. I try to make loads smaller, lighter or easier to grasp	1 2 3 4 5
6. I make sure that my work area is kept clear and uncluttered	1 2 3 4 5
7. I can accept constructive criticism about lifting practices at work	1 2 3 4 5
8. I think through lifting tasks in advance	1 2 3 4 5
9. I am careful to move my feet rather than twist my body	1 2 3 4 5
10. I wear protective clothing, such as gloves, shoes and so on, as appropriate	1 2 3 4 5
11. I read the labels on the outside of boxes to check the weight and contents	1 2 3 4 5
12. If I feel tired or begin to make mistakes, I take a break or change tasks	1 2 3 4 5
13. When lifting objects, I check for the centre of gravity before lifting	1 2 3 4 5

cont'd

14. I feel able to refuse to lift anything that is too heavy for me	1 2 3 4 5
15. I do not risk lifting objects that appear to be unstable or damaged	1 2 3 4 5
16. I pay attention to how things are stacked and observe limits and manual handling safety rules	1 2 3 4 5
17. I take a few moments to stretch and 'warm up' before any lifting operations	1 2 3 4 5
18. I think through lifting movements to reduce carrying as much as possible	1 2 3 4 5
19. I use proper equipment and aids to reach high shelves or stacks	1 2 3 4 5
20. I follow safe lifting techniques	1 2 3 4 5

My score is _____/100 or _____%.

Analysis

Between 80%–100%	Excellent.
Between 60%–80%	Very good – a really high level of manual handling safety awareness.
Between 40%–60%	OK, but there is room for improvement.
Less than 40%	An accident waiting to happen!

Notes

____

____

____

____

PLEASE COMPLETE BEFORE CONTINUING

Chapter 4
Six Steps to Safe Lifting

This chapter shows you how to lift safely and how to avoid risks and injuries when lifting.

Before starting this chapter, please take a few moments to make a note of any ideas or actions in the learning diary and log in Chapter 1.

HOW TO LIFT SAFELY

Here is the simple, but very important sequence to follow when lifting or moving objects by hand. These six steps will ensure that you lift items not only correctly but also in a way which is as safe as possible and which minimizes the risk of injury to yourself and others.

1. **Pause and take position.**
2. **Bend.**
3. **Grip.**
4. **Look.**
5. **Lift with your legs.**
6. **Hold the load close.**

Following these steps will also ensure that you work within the manual handling safety guidelines and regulations that are set down not only by your organization but also by the manual handling safety authorities and regulators.

Safe lifting step 1: POSITION

The very first step – regardless of what you are lifting or its weight – is to take a few seconds to think carefully about how to lift it, the route you are taking when you are carrying it, where you are actually going to put it and, most importantly perhaps, how to position yourself correctly over, around or next to the object. Most injuries and discomfort are caused by an incorrect position that puts undue strain on the muscles or the back.

Step 1: Position

- **Consider what you are lifting.**
- **Consider the best position.**
- **Stand correctly.**
- **Test before lifting.**

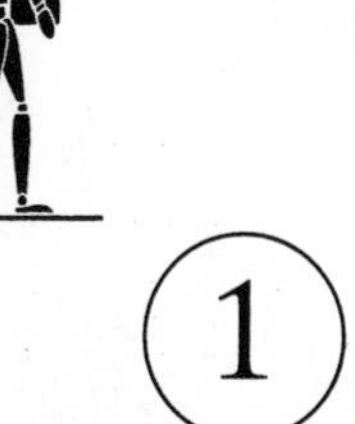

1

The position of your feet is particularly important. They should be placed normal width apart – say about 18 inches – either side of the object so that you have a good solid stance. You should also be as close to the object as possible.

If anything needs to be moved or adjusted or you need to obtain a lifting aid, do this before beginning the operation.

For instance, all of us will occasionally kick a door open whilst carrying an object, and this increases the risk of a manual handling accident or injury. Of course, what we should do is to open the door – or, better still, prop it open – before beginning our lift. Now that you have attended this training programme, performing this kind of action could be deemed as unsafe working practice. This means that, if you did do this, and were involved in a manual handling accident or injury, the liability would be yours because you should have known better.

Safe lifting step 2: BEND

Part of positioning is to make sure you get 'underneath' the load. This involves bending your knees and getting a grip – that is, actually holding the load, or the object – as far as possible beneath the main centre of gravity.

Step 2: Bend

- **Position yourself 'underneath' the load.**
- **Bend with the knees.**
- **Keep your back straight.**
- **Don't stoop or stretch.**

Safe lifting means lifting with your legs, not your back.

Bend your knees, keep your back straight and lower yourself down to the same level as the object you want to lift. Then, when you do lift, you will be able to push up with your legs, keeping your back straight all the time.

Step 3: Grip

- **Put your hands underneath corners, avoiding sharp edges or anything loose.**
- **Test your grip before fully lifting.**
- **Check for rips and tears.**
- **Square up before lifting.**

3

Safe lifting step 3: GRIP

If you are lifting a box or a reasonably square object, as is often the case, place your hands underneath the corners, avoiding any sharp edges or loose materials, and take a firm secure grip.

Before lifting – or certainly taking the

main weight – test your grip fully. Use your grip to lean the weight into your body making sure that its centre of gravity, or the heaviest part of the load, is closest to you.

If you need to wear protective gloves, either to get a better grip or to protect your hands, it is essential that you do this.

You may also need to consider whether you are wearing the correct footwear. Suitable footwear should grip the surface on which you are working and also protect your feet if you drop the load.

Safe lifting step 4: LOOK
When you are moving, carrying or lifting an object, keep looking all around you as you do so. Not only do you need to look where you are going and make sure that you do not walk or bump into things, but also, and perhaps most importantly, it ensures that you lift correctly.

Step 4: Look

- **Keep looking all around as you lift or carry.**
- **Keep your head erect and your back straight.**
- **You steer with your eyes, so look where you are going all the time.**

4

Keep your head upright, your back straight and your shoulders back. If you begin to drop your head then this in turn may cause your back to bend or curve, putting undue strain on the ligaments and muscles of the lower back in particular.

You also steer with your eyes, so look where you are going all the time – not at the floor or where you have been, but straight ahead towards your destination and following the direction of your feet.

Safe lifting step 5: LIFT WITH YOUR LEGS

Step 5: Lift

- **Your legs are the largest muscle group in the body – use them to protect your back which is the most complex and fragile.**
- **Bend your knees and push up.**
- **Test the weight before lifting.**

Your legs comprise the largest muscle group in the body – you only have to look at any weightlifter or power sports athlete to see the truth of this. When a weightlifter or body-builder becomes skilled at lifting very heavy weights, it is usually their legs that become the biggest or strongest body part.

In contrast, our back muscles – and our back structure in particular – are extremely complex and extremely fragile. Backs can be damaged very easily, and with the smallest load. When you take the weight of an object you must do so with your legs.

If you are lifting an object from the floor, rather than bending over and lifting with the arms, it is much safer to crouch down over the object, take a secure grip and then lift up, straightening your knees and pushing up as you do so. When moving or carrying items make sure that the weight is taken by your legs and not by your back.

Safe lifting step 6: HOLD THE LOAD CLOSE

This step is particularly simple and almost deceptively obvious, but is vitally important. The closer you hold the load or the object to yourself the less strain you put on your back. Always make sure that you have the heaviest part of whatever you are lifting towards you and pull it as close to you as possible.

Step 6: Hold close

- **Hold the load as close as possible to your body, to minimize strain on your back.**

6

Often people suffer strains and injuries because they do not want the item which they are lifting to touch their clothing for some reason. For instance, when unloading gardening materials from the boot of their cars, they will deliberately hold them at arm's length and, in so doing, risk painful or disabling injuries as a result.

Lifting: weight guidelines

Whilst the amount that we can actually lift will, of course, vary between individuals and will also depend both on what we are lifting or moving and the conditions or environment in which we are moving it, a set of general, helpful guidelines has been produced.

For instance, when lifting a box from the floor to bench or waist level, it is advised that the weight limit should be within 15–25 kg: the heavier weight should be a secure and compact object that is being held close to the body and the lower weight should be an object that is either larger or bulkier, or is not lifted so closely to the body.

When lifting from the floor or waist level to head height, the weight guideline is between 5–10 kg.

The fact that these weights may seem relatively light compared to the kind of weights that we lift regularly simply highlights the kind of risks to which we expose ourselves everyday when we exceed the manual handling safety guidelines.

Lifting: weight guidelines

- **Floor or waist level to head height: 5–10 kg**
- **Floor or knee level to waist height: 15–25 kg**

  (The higher weight is for loads held close to the body, the lower for loads lifted with arms outstretched.)

Make a note of any points from this section that concern you.

Notes

THE LIFTING SEQUENCE

Practise this technique for a variety of different objects, starting with very light ones.

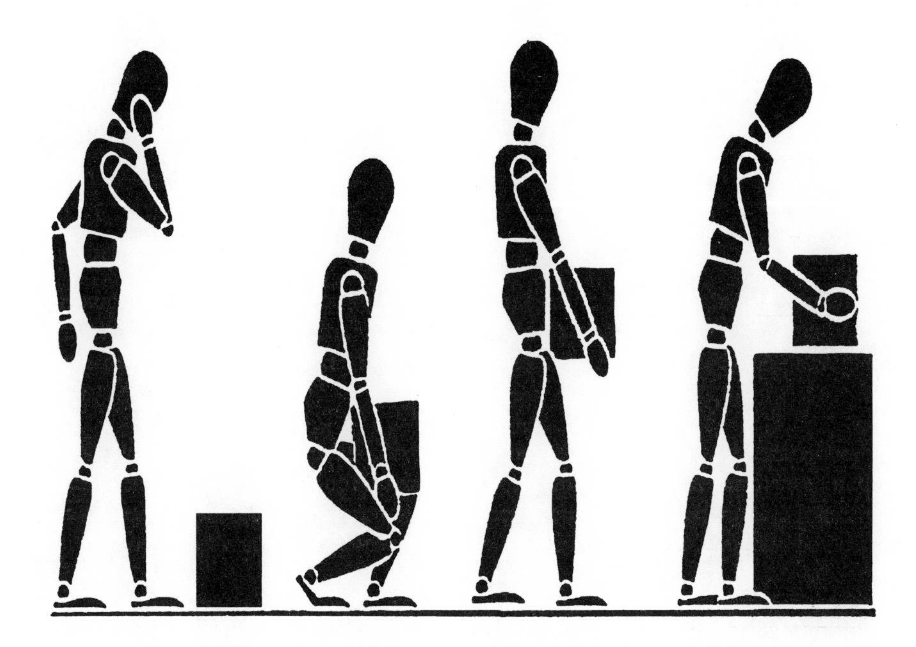

Make a note of any parts of this lifting technique to which you need to pay particular attention.

Notes

Chapter 5
Look After Your Back

This chapter gives you useful guidelines and techniques for avoiding injury and strain to the back.

Before starting this chapter, please take a few moments to make a note of any ideas or actions in the learning diary and log in Chapter 1.

MAKE THE RIGHT MOVES

B *Be aware; your back needs care.*

A *Always bend your knees with practice and ease.*

C *Check you have a clear way; it might save the day.*

K *Kneel on a mat; don't bend your back.*

C *Check the weight first, otherwise you'll curse.*

A *Avoid lifting if you can; it's easy if you plan.*

R *Remain steady and stable, with feet under the table.*

E *Exercise regularly for muscular strength along your whole body length.*

BACK AWARENESS

Stop what you are doing and notice how you are sitting at this moment. Sketch your position or make some notes to describe this.

____

____

____

1.	Is this how you normally sit?	**Yes/No**
2.	Is your spine supported?	**Yes/No**
3.	Are you slouched forward or leaning back?	**Yes/No**
4.	Is your back stiff when waking or in the evening?	**Yes/No**
5.	Do you know of others who also suffer in this way?	**Yes/No**

Top Ten Lifting Errors

The following is a list of the top ten lifting errors identified from our everyday lives and work situations. These top ten errors will be the cause of nearly all manual handling or back-related injuries or disabilities.

1. HOLDING THE LOAD AWAY FROM THE BODY
2. TWISTING
3. STOOPING
4. REACHING UPWARDS
5. EXCESSIVE UP AND DOWN MOVEMENTS
6. CARRYING LOADS FOR LONG DISTANCES
7. STRENUOUS PUSHING OR PULLING
8. UNPREDICTABLE OR UNBALANCED LOADS
9. REPETITIVE HANDLING AND LIFTING
10. TOO STRENUOUS WORK-RATE

PLEASE COMPLETE BEFORE CONTINUING

LIFTING CHECKLIST

Before starting any manual handling tasks consider the following checklist:

✓

- ☐ Do I really need to move this?
- ☐ Is there a piece of equipment to help me?
- ☐ Is it too heavy?
- ☐ I am fresh and feeling able to lift?
- ☐ Is my route clear?
- ☐ Hold the load as close to the body as possible.
- ☐ Don't twist, stretch or stoop.
- ☐ Don't reach or stretch upwards – use steps.
- ☐ Avoid excessive or repeated movements.
- ☐ Don't carry loads over long distances.
- ☐ Don't push or drag very heavy objects.
- ☐ Lift unbalanced objects carefully.
- ☐ Take regular breaks when carrying out repetitive tasks.

YOUR ATTITUDE TOWARDS LIFTING MAKES THE DIFFERENCE!

Manual handling accidents do not just happen; something causes them, and that something is usually people or, to be more specific, people's attitudes. So, how do we define 'attitude' or, more importantly, a 'safe attitude'?

Here are four keys to having a positive attitude towards safety.

1. Be informed

Take time to understand and think about the hazards and risks that exist, not only in your own work environment but also in the various tasks and jobs you undertake during the day.

2. Be aware of safety procedures

Every organization has safety rules and procedures, including yours. It is very important that you learn the rules concerning health and safety and then keep to them. Not only is it clearly unsafe to break safety policies and procedures, you may actually be breaking the law and therefore be liable to prosecution.

3. Cooperate with safety representatives

The Health and Safety at Work Act states that it is the responsibility of every employee to cooperate fully with safety representatives and managers who are implementing safety policy and procedures. To put it more positively, you will benefit most from good safety procedures, so it makes sense to cooperate with other people around you who are working towards that goal.

4. Be alert

Accidents happen when we walk around with our blinkers on, so always be alert, awake and aware of what is going on around you and of what hazards may exist, or potentially exist, in all situations.

Make a note of any points from this section that concern you.

Notes

COMMON ATTITUDES THAT CAUSE MANUAL HANDLING ACCIDENTS AND INJURIES

Now that we have considered some positive attitudes that can help us prevent manual handling accidents and injuries, let us look at the more negative attitudes that can actually cause manual handling accidents in the first place.

1. Overconfidence

Overconfidence means thinking that manual handling accidents cannot happen to us, that they only happen to other people and believing that we are 'too clever' or 'too good' at what we do.

2. Laziness

There is a saying that states, 'A short-cut is only a fast route to a shortcoming'. Do not try to cut corners or take unnecessary risks. You may be risking your own life as well as someone else's.

3. Stubbornness

Many jobs require people to wear personal protective equipment (PPE), such as hard hats, ear protection, special clothing, safety boots and so on. In fact, we all wear a particular form of protective equipment each time we make a journey by car – our seat belts.

However, consider how many people wore a seat belt for every journey before it was made law. Statistics show that it was very few of us, and the reason was probably stubbornness and bad habits. Even though all of us understood that it was better to wear a seat belt than to fly through the windscreen in the event of an accident, it took an Act of Parliament to make people do so for every journey.

4. Impatience

Many manual handling accidents are caused by people trying to do a task too quickly without paying due care and attention to what is going on around them or the consequences of not carrying it out properly. Very often we end up having to do something twice because we did not do it properly the first time. As the old saying goes: 'More haste, less speed'.

5. Ignorance

Clearly, if someone is ignorant of the dangers inherent in any particular operation or task, it might be unfair of us to expect them to know how to behave safely. However, the law says that ignorance is no defence.

When a Health and Safety Inspector investigates a company after an accident or injury, the company must prove that its employees were sufficiently trained and had the right knowledge and skills for the jobs and operations that they were carrying out.

Ignorance is no excuse when it comes to health and safety, so make sure that you really understand and know the hazards and risks that exist in your workplace.

6. Showing off

A moment's levity can lead to long-term regret. We have all occasionally done something foolish and, fortunately, probably escaped too serious a consequence. However, always be aware that when we let down our guard we increase the risks of manual handling accidents.

7. Forgetfulness

We are all probably guilty of this from time to time. Even when we know that we should be doing, even with our normal best intentions, occasionally we forget. Safety is a full-time job, and it requires your full-time attention. Always consider safety; never forget its importance.

Make a note of any points from this section that concern you.

Notes

___

___

___

___

___

___

___

Chapter 6
Learning Review

This chapter looks at practical ways of transferring key learning points to your own work situation.

Each exercise must be completed fully and reviewed with your manager, colleagues, safety representative or work colleagues.

Before starting this chapter, please take a few moments to make a note of any ideas or actions in the learning diary and log in Chapter 1.

TEST YOUR KNOWLEDGE (1)

1. How many working days are lost every year due to manual handling-related accidents, injuries or disorders?

2. Name three high-risk manual handling activities.

   1.

   2.

   3.

3. What are the six steps for preventing back injuries and manual handling disorders?

   1.

   2.

   3.

   4.

   5.

   6.

4. What does RSI stand for?

5. What is the name of the regulations which govern the lifting, loading or carrying of objects?

6. What is the manual handling safety weight guideline for lifting an item from the floor to knee or waist level?

PLEASE COMPLETE BEFORE CONTINUING

TEST YOUR KNOWLEDGE (2)

1. List three things to consider before attempting to lift an object or perform a series of repetitive lifting tasks.

   1.

   2.

   3.

2. List three possible injuries that might occur or be caused by incorrect or improper lifting.

   1.

   2.

   3.

3. List three things to consider when lifting heavy objects.

   1.

   2.

   3.

4. What percentage of reported manual handling accidents are caused by manual handling or lifting?

   a) 10% b) 20% c) 30%

5. On average, how many days does each injury result in the person being off work?

   a) 5 days b) 10 days c) 20 days

6. As well as back injuries, list two other injuries that are often caused by manual handling lifting accidents.

   1.

   2.

PLEASE COMPLETE BEFORE CONTINUING

Test Your Knowledge (3)

1. Describe how you would approach lifting a large heavy object, when there are two of you to lift it.

2. Describe how you would lift something on to a high shelf.

3. Describe how you would lift a heavy bucket or tin of paint.

4. Give some instructions on how to use a set of wheeled trucks.

5. List three ways that you could reduce the **risk** of injury from lifting.

   1.

   2.

   3.

Please complete before continuing

TEST YOUR KNOWLEDGE (4)

1. Getting a good grip means:

   a) not wearing gloves

   b) hooking your hands under the load

   c) holding the load against your body

   d) sliding objects close to you when on a shelf or desk

   e) moving one object at a time, not several piled or tied together

   (CIRCLE ALL THAT APPLY)

2. Many manual handling injuries are caused by using incorrect techniques to handle light weights over a period of time rather than single incidents.
   TRUE/FALSE

3. Your employer must avoid or remove the need for manual handling or lifting wherever reasonably practical.
   TRUE/FALSE

4. Your employer or manager must assess each type of lifting operation that can't be avoided for likely hazards and risk of injury.
   TRUE/FALSE

5. As an employee, you must:

   a) follow appropriate systems of work at all times

   b) use your own common sense in assessing safety

   c) develop your own system of working safely

   (CIRCLE ALL THAT APPLY)

6. As an employee, you must:

   a) make proper use of lifting aids and equipment

   b) use a lifting aid only if you think it is needed

   c) use lifting aids and equipment only if you are trained

   (CIRCLE ALL THAT APPLY)

PLEASE COMPLETE BEFORE CONTINUING

Case Study

Read the following case study and then answer the questions on the following page.

Personnel Accident Report

'The warehouse was a real mess. Old boxes were left lying around – pallets full of goods awaiting collection left right in the way of the gangway, and bits and pieces of things waiting to be sorted left all over the place. Dave, the supervisor, was supposed to be there to help that afternoon, but he had to go to the other branch to sort out some problems. Mondays are also really busy days. We have extra collections all morning from the weekend and then our weekly delivery in afternoon. Business has been really good over the last few weeks, and that afternoon's load was probably much larger than normal.

'I started unloading with Nigel, but he had to go and see to some more collections, so I continued on my own. The drivers are never much help, and just sit and drink tea and moan about hitting all the traffic if you don't unload quickly.

'I had been using the trucks most of the time, but then one of the wheels started to jam and eventually I couldn't use it at all, so I had to carry everything by hand, and some of the boxes are really heavy. I had almost finished when I came to the final pallets of boxes. I reached up to lift up the top one, and that's when it happened. It felt like someone had just put a knife in the base of my back. The pain was unbelievable. I just couldn't stand up or anything. I shouted for help, and Nigel went off to get Sue from Accounts, who knows about first aid. She called the ambulance straightaway.

'I was off for four weeks, and now have to wear a back support whilst at work. I haven't been able to play football since it happened. (I used to play Sunday league every week.)

'My doctor said that I should have said something about the state of the warehouse to somebody before, but I didn't know that this was going to happen.'

Case Study: Questions

1. **What do you think caused the manual handling accident?**

2. **Who was at fault?**

3. **Should the person in question have mentioned the state of the equipment and done something about it themselves?**

4. **What would you do now?**

Please complete before continuing

MANUAL HANDLING CHECKLIST

Make a list of everything that you might consider when assessing manual lifting tasks or procedures. You might find it useful to make a list under each of these four categories.

1. The task itself	**2. The load**
3. The work environment	**4. The individual**

PLEASE COMPLETE BEFORE CONTINUING

Reducing Risks

Make some notes on the following.

1. How can you avoid the need for manual handling or lifting altogether?

2. In your organization, how well have people been trained in manual handling techniques?

3. How could this be further improved?

4. Are people able to refuse to lift or shift heavy or unsafe loads?

5. Is the working environment safe and properly organized? How could this be improved?

Please complete before continuing

SAFE LIFTING IDEAS

Tick what you consider to be the five best pieces of advice.

☐ Always consider avoiding manual handling wherever possible.

☐ If you see someone lifting, carrying or loading in a way that is not particularly safety-aware, point it out to them.

☐ Assess the risk of each handling operation.

☐ Use mechanical equipment wherever possible.

☐ Try to make loads smaller, lighter or easier to grasp.

☐ Make sure that your work area is kept clear.

☐ Accept constructive criticism about lifting practices.

☐ Move your feet rather than twist your body.

☐ Wear protective clothing, such as gloves, shoes and so on – as appropriate.

☐ Read the labels on boxes to check the weight and contents.

☐ Take a regular break or change tasks.

☐ Check for the centre of gravity before lifting.

☐ Refuse to lift objects that are too heavy.

☐ Do not risk lifting objects that appear to be unstable.

☐ Pay attention to how items are stacked.

☐ Observe limits and rules.

☐ Stretch and 'warm up' before any lifting operations.

☐ Think through lifting movements to reduce carrying.

☐ Use proper equipment and aids to reach high shelves.

☐ Follow safe lifting techniques.

PLEASE COMPLETE BEFORE CONTINUING

Appendix Suggested Answers to the Knowledge Tests

Test your knowledge (1): suggested answers

1. 5.5 million

2. Heavy objects
   Awkward objects
   Dirty or wet objects

3. Pause
   Bend
   Grip
   Look
   Lift
   Hold

4. Repetitive strain injury

5. Manual Handling Operations Regulations 1992

6. 15–25 kg

Test your knowledge (2): suggested answers

1. Weight, size, shape and condition of object
   Plan the activity
   Consider wearing protective clothing and using lifting aids

2. Back strain
   Slipped disc
   Pulled muscle

3. Position of your feet
   Getting a sure grip
   Lifting with the knees, and keeping a straight back

4. c) 30%

5. c) 20 days

6. Hernia
   Shoulder injuries
   Damaged toes and feet from dropping heavy objects.

Test your knowledge (3): suggested answers

1. Work with someone of a similar height and strength
   Designate one person as the 'leader' and call commands
   Lift in unison and from the hips/knees
   Move smoothly and thoughtfully

2. Make sure that it is not TOO heavy to store lower down
   Use appropriate steps/equipment
   Seek help/assistance
   Perform the lift in two stages, to avoid twisting, and check grip

3. Get a good stance with feet either side of the item to be lifted
   Take a firm grip and test the weight
   Lean over the load
   Look ahead and up

4. Don't overload
   Keep your back straight
   Transfer the weight on to wheels fully before trying to move
   Push smoothly and carefully; avoid twisting

5. Avoid making the lift at all by using a mechanical aid or reorganizing work
   Use the correct techniques
   Work in a clear, well organized environment or area

Test your knowledge (4): suggested answers

1. b) Hooking your hands under the load
   c) Holding the load against your body
   d) Sliding objects close to you when on a shelf or desk
   e) Moving one object at a time, not several piled or tied together

2. True

3. True

4. True

5. a) Follow appropriate systems of work at all times
   b) Use your own common sense in assessing safety
   c) Develop your own system of working safely

6. a) Make proper use of lifting aids and equipment
   c) Use lifting aids and equipment only if you are trained

Case study

The purpose of the case study exercise is for students to apply safety knowledge and awareness.

Whilst there are no 'right' answers, students should highlight legal regulations and standards that have been broken and practical ways of enforcing these.

Lessons learnt from each situation should also be identified, both in terms of what caused the incidents and what could prevent them from happening again in the future.

Manual handling checklist, reducing risks and safe lifting ideas

These exercises are designed to encourage the student to think carefully about manual handling and how to reduce the risks and, as such, have no 'right' answers. They may, however, be kept as notes for future reference and amended or added to in light of experience at work or simply used as a reminder of safe practice in manual handling.

HEALTH AND SAFETY WORKBOOKS

The 10 workbooks in the series are:

Fire Safety	0 566 08059 1
Safety for Managers	0 566 08060 5
Personal Protective Equipment	0 566 08061 3
Safe Manual Handling	0 566 08062 1
Environmental Awareness	0 566 08063 X
Display Screen Equipment	0 566 08064 8
Hazardous Substances	0 566 08065 6
Risk Assessment	0 566 08066 4
Safety at Work	0 566 08067 2
Office Safety	0 566 08068 0

Complete sets of all 10 workbooks are available as are multiple copies of each single title. In each case, 10 titles or 10 copies (or multiples of the same) may be purchased for the price of eight.

Print or photocopy masters

A complete set of print or photocopy masters for all 10 workbooks is available with a licence for reproducing the materials for use within your organization.

Customized editions

Customized or badged editions of all 10 workbooks, tailored to the needs of your organization and the house-style of your learning resources, are also available.

For further details of complete sets, multiple copies, photocopy/print masters or customized editions please contact Richard Dowling in the Gower Customer Service Department on 01252 317700.

YSTRAD MYNACH COLLEGE
LIBRARY